The Automotive Evolution

Blockchain Technology in the Transportation Industry

Table of Contents

1. Introduction 1

2. The Road to Technological Integration 2

    2.1. The Internet of Things and Connected Vehicles 2

    2.2. Artificial Intelligence and Autonomous Driving 3

    2.3. Integration of Blockchain Technology 3

3. The Genesis of Blockchain: A Primer 5

    3.1. Blockchain: The Basics 5

    3.2. Emergence of Blockchain 6

    3.3. Bitcoin: The First Use Case of Blockchain 6

    3.4. From Bitcoin to a Broad Application Spectrum 6

    3.5. Understanding Smart Contracts 7

    3.6. The Scalability Challenge 7

    3.7. Blockchain and Automotive: Gearing for the Future 8

4. Linking Blocks: Understanding Blockchain Functionality 9

    4.1. Block and Chain: Explaining the Basics 9

    4.2. Decentralization: The Backbone of Blockchain 10

    4.3. Blockchain Types: Public and Private 10

    4.4. Where Blockchain Excels 11

5. Blockchain and Automotive: An Unlikely Match, or A
Revolution in Disguise? 13

    5.1. A Primer on Blockchain Technology 13

    5.2. Blockchain: The Unusual Suspect for the Automotive Sector . 13

    5.3. On the Road to Authenticity with Blockchain 14

    5.4. Blockchain as Key to Safer Mobility Services 14

    5.5. Driving Efficiency with Blockchain in Automotive 15

    5.6. Ensuring Data Privacy with Blockchain 15

6. Evolving Supply Chains: How Blockchain Changes the Game 17

    6.1. Changing Lanes with Blockchain 17

6.2. Recalibrating Processes: Efficiency and Speed 18

6.3. Unmanned Roadblocks: Enhanced Security and
Authenticity 18

6.4. Predictive Maintenance: Driving Proactivity with Data 19

6.5. Clearing the Windshield: Transparency and Trust 19

7. Driving Authenticity: Blockchain in Vehicle History Verification . 21

7.1. The Significance of Vehicle History in the Automotive
Industry 21

7.2. Blockchain: An Overview 22

7.3. Application of Blockchain in Vehicle History Verification 22

7.4. Real-World Examples of Blockchain in Action 23

7.5. The Way Forward 23

8. Fueling Transparency: Payment Systems & Leasing 25

8.1. Blockchain and Payment Systems in Automotive 25

8.2. Blockchain Revolutionizing Leasing 26

8.3. Encouraging Transparency and Efficiency 27

8.4. Final Thoughts 27

9. Road of the Future: Blockchain in Traffic Management and
Autonomous Vehicles 28

9.1. Integrated Traffic Management Systems 28

9.2. The Role of Decentralized Infrastructure 29

9.3. Exploring Blockchain's Potential 29

9.4. Roadmap to Autonomous Vehicles 30

9.5. In conclusion 31

10. Bumps Ahead: Challenges in Implementing Blockchain 32

10.1. Clearing the Fog: Understanding Blockchain 32

10.2. Trust: The Invisible Battleground 33

10.3. Regulatory Roadblocks and Standardization Jams 33

10.4. Crumbling Infrastructure Worries 34

10.5. Skills Shortage: A Speed Limit? 34

11. Looking Through the Windshield: The Future of Blockchain in the Automotive Industry 36

    11.1. The Underlying Technology: A Simplified Overview 36

    11.2. Blockchain in Automotive: Current Applications 37

    11.3. The Road Ahead: Future Applications 37

    11.4. Challenges and Framework for Implementation 38

Chapter 1. Introduction

In a world that is becoming inextricably intertwined with technology, even traditional industries such as automotive are not left untouched. Welcome to "The Automotive Evolution: Blockchain Technology in the Transportation Industry", a Special Report that teases apart this intricate fusion of technology and transportation. This isn't just another high-tech jargon-loaded document, but rather a down-to-earth unraveling of how blockchain technology, a term often used and seldom understood, is charting a new course for the automotive industry. Let's journey together on this enlightening expedition, which promises to illuminate, inform, but most importantly, transcend the complexities of technology. Your understanding of the automotive sector is about to shift gears. If you thought the high-speed chase for innovation ended with electric vehicles and self-driving cars, buckle up; you're in for a thrilling ride!

Chapter 2. The Road to Technological Integration

As the landscape of the automotive industry is redrawn continually in the wake of emerging technologies, there has been an exhaustive pursuit of integration, not just of different vehicular systems internally but also with external technological modules. Technological integration on the road is a vast subject that encompasses a myriad of individual topics such as Internet of Things (IoT), artificial intelligence (AI), and blockchain.

2.1. The Internet of Things and Connected Vehicles

The rise of the Internet of Things (IoT) revolutionized the level of interconnectivity between various systems, and vehicles were no exception. Connected vehicles equipped with IoT devices can directly communicate with their surroundings. This direct communication technology, or vehicle-to-everything (V2X), enables vehicles to connect to anything, including infrastructure, networks, and other vehicles. It provides real-time information to drivers, thereby increasing safety, reducing traffic congestion, and conserving energy.

However, the current complexity of this technology and the high cost of large-scale implementation inhibit its universal adoption. Car manufacturers and IoT developers are working closely to cut costs and simplify the processes, making connected vehicles an increasingly common sight on global roads.

2.2. Artificial Intelligence and Autonomous Driving

Artificial Intelligence (AI) is another technology that is increasingly integrated into the vehicle industry, predominantly in autonomous or self-driving vehicles. AI platforms are utilized extensively to simulate the decision-making process of the human brain while driving and react to different scenarios on the road.

Machine learning – a subset of AI, allows the autonomous vehicles to learn from the data collected during each ride and continually improve their performance. Millions of miles have been logged by autonomous vehicles for reinforcement learning to improve safety and reliability. Nevertheless, concerns about the ethical implications of AI and public acceptance of autonomous vehicles still pose obstacles to widespread adoption of this technology.

2.3. Integration of Blockchain Technology

Blockchain, the technology that brought us cryptocurrencies like Bitcoin, also has massive potential for revamping the transportation industry. With blockchain, all transactions and exchanges of data can be recorded securely and transparently on a decentralized ledger that is almost impossible to tamper with.

In the context of the automotive industry, blockchain can manifest several applications. For instance, it can streamline supply chain by maintaining a comprehensive, immutable record of parts from the point of manufacture to installation. This promises to eliminate counterfeiting, enhance safety, and foster consumer trust.

Blockchain can also be the backbone of a secure digital ecosystem for connected cars, where data can be exchanged safely between vehicles, infrastructure, and third parties. Mobility services like ride-sharing could massively benefit from blockchain by eliminating intermediaries and directly connecting drivers with passengers.

Moreover, blockchain could transform vehicle lifecycle management by keeping an accurate and comprehensive history of the vehicle, making second-hand market more transparent and trustworthy.

In conclusion, the integration of IoT, AI and blockchain into the automotive industry is not a straightforward venture. Significant challenges in terms of costs, scalability, public acceptance and regulatory frameworks are yet to be overcome. The road to technological integration is a grueling, uphill journey, but with the immense benefits in terms of safety, efficiency, sustainability, and customer satisfaction, the automotive industry is all set to accelerate towards this direction.

Chapter 3. The Genesis of Blockchain: A Primer

From its humble, almost obscure beginnings, blockchain rose to fame as the underpinning technology of Bitcoin, the highly publicized and widely-discussed digital cryptocurrency. Many, intrinsically linking blockchain with Bitcoin, viewed it solely as a facilitator for these emerging digital currencies. However, the inherent characteristics of blockchain technology soon began to resonate with a broader spectrum of industries, the automotive sector being one of them. Blockchain and its relevance for automotive and transport aren't an obvious pairing at first glance, but digging deeper, the potential becomes evident.

3.1. Blockchain: The Basics

At its core, a blockchain is a data structure that maintains a record of transactions distributed across several computers in a network. This public digital ledger ensures that each transaction is authentic and has been confirmed by the majority of participating nodes in the network. The verifying nodes cross-check the validity of each transaction, which subsequently becomes part of the blockchain - a securely chained sequence of blocks, hence the name blockchain.

Blockchain's key characteristics include decentralization, transparency, permanence, and security. These strengths stem from cryptography, the bedrock of blockchain. Cryptography ensures that each transaction is securely encrypted, maintaining privacy while preventing tampering. Such strict adherence to security, coupled with transparency and immutability, kindles trust amongst participants in the blockchain network, even in the absence of a centralized authority.

3.2. Emergence of Blockchain

The inception of blockchain dates back to 2008 when an individual or group of individuals under the pseudonym Satoshi Nakamoto published a white paper titled "Bitcoin: A Peer-to-Peer Electronic Cash System". Nakamoto's model proposed a network where transactions could be verified without the need for a trusted third party, thus envisioning a genuinely decentralized system. Despite the pseudonymous author's primary focus on digital currency, the distributed ledger technology proposed had implications far beyond Bitcoin alone, with potential for widespread application in various sectors.

3.3. Bitcoin: The First Use Case of Blockchain

Bitcoin, the first cryptocurrency, is a vivid example of blockchain's first major application. It demonstrated the efficacy of a peer-to-peer network where each participant or 'node' could verify transactions. Bitcoin achieved disruptive fame not merely as a volatile digital currency, but also as an emblem of the power of decentralization - the cornerstone principle of blockchain.

In the Bitcoin network, participants known as miners solve complex mathematical problems to verify and add transactions to the blockchain. Miners receive Bitcoin as an incentive, ensuring continuous network participation and transaction validation. This system elegantly balances rewarding effort with promoting honesty.

3.4. From Bitcoin to a Broad Application Spectrum

As insightful individuals and businesses began to see beyond Bitcoin, they discerned the potential of the underlying blockchain technology. Decentralization, transparency, security, and immutability are valuable traits in any transaction network, and industries saw the potential applications towards establishing trust, verifying transactions, recording data, and ensuring traceability.

Automotive and transportation industries, realizing the intrinsic value and broad array of applications of blockchain, began exploring the technology as a solution to various industry-specific challenges. These include traceability in supply chains, managing vehicle lifecycles, confirming vehicle service history, and ensuring the authenticity of parts, among others.

3.5. Understanding Smart Contracts

As blockchain technology was studied further, the concept of smart contracts emerged as an extension of the core principles. Conceived by cryptographer and legal scholar Nick Szabo in the early '90s, smart contracts encoded in a blockchain enable automatic, decentralized execution of contract terms when pre-set conditions are met.

This innovation, albeit requiring a paradigm shift for regulatory and control mechanisms traditionally vested in centralized authorities, allows for tremendous efficiency and acceleration in transactional processing. In the automotive industry, smart contracts are being explored for traceability in supply chains, vehicle leasing and service contracts, and even for payment of vehicle-related taxes and fees.

3.6. The Scalability Challenge

However, like any nascent technology, blockchain too faces a fair share of challenges. One prominent issue is scalability. The computational power required to verify transactions and add blocks to a blockchain grows with the network, particularly in proof-of-work systems like the Bitcoin network. As the energy and latency requirements increase, so does the challenge of scaling the blockchain effectively.

Nevertheless, researchers and developers worldwide are persistently working on blockchain scalability solutions, such as proof-of-stake protocols, sidechains, and sharding techniques, aiming to address this inhibiting factor.

3.7. Blockchain and Automotive: Gearing for the Future

Looking beyond these challenges, what seems certain is that blockchain technology holds immense potential for automotive and transportation sectors. Blockchain's core principles align well with the industry's needs for robust security, transparency, and fraudulent behavior mitigation, signaling a promising future for this burgeoning relationship.

Driven by tangible blockchain-powered advantages, automotive industry stakeholders worldwide are investing time, resources, and capital into research and development of blockchain solutions. Some trailblazing companies have already begun implementing blockchain-powered applications, affirming the potential this still young but powerful technology embodies.

As this overview has shown, blockchain's genesis places it firmly in the high-tech arena. However, it holds the potential to profoundly impact sectors far removed from its birthplace. The automotive sector is one such industry touching tangibly upon blockchain's specialties, ready to harness its utility to drive the next significant era of innovations and transformations. This exploration marks merely the beginning of the road in deciphering how embedding blockchain in the automotive and transportation industries could redefine the trajectory for their technological evolution.

Chapter 4. Linking Blocks: Understanding Blockchain Functionality

To truly comprehend the transformative power of blockchain technology in the automotive industry, we must first unravel what it is and how it works. At the center of its functionality is the idea that it combines power, tech, and security in an unbreakable blend, offering industries a new way to store, protect, and transfer everything from financial transactions to digital identities.

4.1. Block and Chain: Explaining the Basics

Blockchain, as its name suggests, involves two core components: blocks and chains. Every 'block' in a blockchain network hosts a comprehensive record of transactions or data, recorded in digital form. It's here where it all begins - transactions are first encrypted and then recorded in a block. And this is no simple undertaking. Ensuring such a transaction is credible requires what is commonly known as 'proof of work', a complex mathematical problem diligently resolved by powerful computers, called nodes, that belong to the network.

Once the 'proof of work' is established, the block having the bundled transactions gets added to the blockchain. The 'chain' part of a blockchain, refers to the way these blocks are chronologically linked, creating a cohesive and irreversible line of transaction history. Each block is connected to its predecessor by a unique code known as a hash - a fingerprint of sorts derived mathematically from the data in the previous block. This inescapable connection between blocks makes the blockchain resistant to manipulation, ensuring its reputation as a secure and reliable ledger of transactions.

4.2. Decentralization: The Backbone of Blockchain

In traditional centralized networks, data is stored in a single hub, like a bank or a government database. In contrast, a blockchain is a decentralized and distributed digital ledger. In essence, every computer that participates in the blockchain network, also known as a node, has a copy of the complete blockchain. Every node has full transparency and authority to validate transactions.

Once a block is added to the blockchain, its data becomes unalterable. This immutability derives from the distributed nature of the blockchain: altering one copy of the blockchain would mean having to alter every single copy across all nodes, an effort so computationally immense, it's practically impossible. This grants the blockchain its badge of security and trust.

Decentralization also provides an additional level of resilience and fault tolerance - the system remains operable even if some nodes fail or are hacked, as the remaining nodes preserve the integrity of the blockchain.

4.3. Blockchain Types: Public and Private

Depending on the accessibility and who has the permission to validate transactions, blockchains are divided into 'Public' and 'Private'. As the name suggests, public blockchains are decentralized and open to anyone in the network. Bitcoin, the most famous implementation of blockchain technology, is an example of a public blockchain.

On the other hand, private blockchains, also known as consortium or enterprise blockchains, operate within an organization or among a specific group where members are pre-determined. Permission to validate transactions, read or write to the blockchain is purely subjected to the access control mechanism set by the network administrators. This makes private blockchains a great choice for businesses where confidentiality of data and transactions is paramount.

4.4. Where Blockchain Excels

The key characteristics of blockchain - decentralization, security, transparency, immutability, traceability, and automation - make it highly applicable in many areas. The financial sector was the first to leverage blockchain skillfully, where it efficiently streamlined transactions. It also proved to be a game-changer in the field of supply chain management, by providing end-to-end product visibility, traceability, and reducing counterfeit goods.

Beyond these realms lies the vast auto industry, an untapped sector that presents an ocean of possibilities for blockchain. Here, smart contracts□an integral part of blockchain technology□could revolutionize workflows by replacing traditional contracts, adding automation, and minimizing the need for intermediaries.

In the world of connected and autonomous cars, blockchain can provide a secure framework, offering a reliable environment for the sharing and storing of vehicle data and digital identities. Furthermore, a transparent, unchangeable ledger would have a significant impact on vehicle service history, recalling issues, and even car finance. Therefore, understanding the basics and functionality of blockchain becomes instrumental in recognizing its potential effect on the transportation industry.

In conclusion, blockchain appears to be more than just a technological disruption. This ingenious merging of cryptography, decentralization, and consensus algorithms holds within its grasp the power to transfigure the operations in the political, social, and economic landscapes. The potential for a seismic shift in the automotive industry is palpable, and as we continue our journey, we will delve deeper into how blockchain could rev the auto industry into the fast lane of innovation.

Chapter 5. Blockchain and Automotive: An Unlikely Match, or A Revolution in Disguise?

Blockchain technology, once considered a niche concept wrapped in mystery, has transcended its initial association solely with cryptocurrency. Today, this technology fuels numerous technological innovations, including in the automotive sector.

5.1. A Primer on Blockchain Technology

At its core, blockchain is a decentralized and distributed digital ledger. Its inherent transparency and immutability make it an exceptional tool for data validation. It is composed of a series of blocks, each containing an aggregation of records, known as transactions. When a new transaction occurs, it is verified across the entire network and then added to a block. Once a block is filled with transactions, it is time-stamped and linked to the previous block in the chain, hence forming a blockchain. The transactions stored in a blockchain cannot be tampered or altered without consensus from the majority of the network participants, thus the data stored holds an unmatched degree of trustworthiness.

5.2. Blockchain: The Unusual Suspect for the Automotive Sector

So, how does this tie into the automotive industry? The sector is undergoing a shift in paradigms, transitioning from owning vehicles to leveraging mobility services. Besides, the emergence of trends like autonomous vehicles, electric vehicles, connected cars, and the Internet of Things (IoT) is adding new dimensions to the industry. Amidst these changes, the integration of blockchain is the unlikely solution to challenges such as data privacy, transparency, efficient operations, and security.

5.3. On the Road to Authenticity with Blockchain

One of the significant areas of application of blockchain in automotive is ensuring the authenticity and transparency of vehicle history. It can help authenticate everything from vehicle ownership, mileage, accident history to its maintenance records. Besides, it also contributes to keep track of parts and components, reducing the chances for counterfeit parts to enter the supply chain. This authenticity decreases fraud, ensures safe transportation systems, provides buyers with accurate history and increases the overall integrity of the automotive ecosystem.

5.4. Blockchain as Key to Safer Mobility Services

Research indicates that with the rise in autonomous and shared vehicles, an extensive array of mobility services would be needed. Here again, blockchain can play a pivotal role. Through the use of smart contracts, blockchain ensures the secure sharing of resources and facilitates automated micropayments. The harnessing of blockchain technology, integrated with IoT, can take vehicle leasing, maintenance and usage services to the next level.

5.5. Driving Efficiency with Blockchain in Automotive

The automotive industry typically operates with vast supply chains and numerous legalities. The merging of blockchain technology can streamline these operations and ensure efficient workflows. For instance, a blockchain can track and store information about logistics, financial transactions, compliance records and agreements. Furthermore, blockchain's immutable nature means that these details cannot be altered, reducing the chances of disputes and litigation.

5.6. Ensuring Data Privacy with Blockchain

Connected and autonomous vehicles generate massive amounts of data, which creates the risk of invasion of privacy and data thefts. Blockchain provides a solution also in this space. The data can be stored on encrypted blocks within the blockchain, ensuring its integrity and security. Through its decentralization, a blockchain essentially empowers users over their data, giving them control on access and utilization, ensuring data privacy to an extent unimaginable before.

As we traverse further down the road of the automotive industry's evolution, blockchain's potential in reshaping the sector becomes increasingly apparent. An unlikely match at first glance, the fusion of these two worlds reveals itself to be a powerful combination of innovation, efficiency, and authenticity – a revolution truly in disguise. While the ride may be bumpy as the industry navigates this new landscape, one thing is clear: blockchain technology is quickly becoming an essential part of the automotive industry's journey towards accessible, efficient, and secure transportation systems. The future of automobiles and blockchain is intertwined more deeply than we could've ever imagined, and thus this unlikely match is becoming a revolutionary pairing, ushering in the new wave of automotive industry evolution.

Chapter 6. Evolving Supply Chains: How Blockchain Changes the Game

Popular wisdom holds that the combustion engine was the linchpin around which the automotive industry evolved. Here's an invitation to challenge that thinking. Picture the supply chain - connecting raw materials to manufacturers, manufacturers to dealers, and finally, bringing a brand new car to your doorstep. Remarkably high standard processes, yes, but also, pervasive inefficiency, vulnerability to fraud, and obscurity. Enter blockchain - the phoenix promising to turn this game on its head.

6.1. Changing Lanes with Blockchain

Imagine a world with fully transparent, seamlessly efficient, and extensively secure automotive supply chains. This is not a utopian fantasy but a reality brought forth by the incorporation of blockchain. In the context of supply chains, blockchain technology offers three pivotal advantages:

1. Transparency: Every transaction is recorded on a blockchain, maintaining a real-time, comprehensive account of the product's journey from raw material to end-user.

2. Efficiency: Enabling direct, peer-to-peer interaction eradicates the countless middlemen currently convoluting processes and slowing down transactions.

3. Security: The decentralized nature of blockchain and cryptographic sealing of blocks make tampering almost impossible, securing against fraud and counterfeit products.

6.2. Recalibrating Processes: Efficiency and Speed

In the labyrinth of processes that constitute a supply chain, confusion and delay often prevail. Operational inefficiencies lead to cost escalations, while a lack of real-time traceability hampers effective response to recalls and other issues. Blockchain, a decentralized ledger, ensures the record of every transaction: a testament set in digital stone, accessible to all relevant parties: manufacturers, suppliers, carriers, dealers, and consumers.

Take delivery timelines for example. Typically, a complex web of interactions involving manufacturers, brokers, freight forwarders, and shipping carriers contributes to inefficiency and incorrect delivery estimations. With blockchain's transparent ledger, the entire transportation journey of a part or vehicle is visible, from the manufacturer's conveyor belt to the customer's driveway. It doesn't just stop there. The ripple effects of this increased visibility leads to improved inventory management, curtailing excess or insufficient inventory.

6.3. Unmanned Roadblocks: Enhanced Security and Authenticity

Security is a leading concern in today's automotive ecosystem. Blockchain's immutable nature makes it resilient to fraudulent interjections and modifications. With cryptographic sealing of blocks, altering recorded transactions involves extensive computational effort, discouraging would-be hackers.

Counterfeit auto parts constitute an expanding problem in the global automotive industry. They pose significant safety risks and tarnish the reputation of auto companies. By recording each material or part's history, including its manufacturing, shipment, and installation details, blockchain makes it nearly impossible for counterfeit parts to infiltrate supply chains.

Moreover, this immutable record of transactions assists in reinforcing warranty and insurance claims. Instead of losing themselves in a paperwork labyrinth, customers can easily and unambiguously verify their claims using the blockchain.

6.4. Predictive Maintenance: Driving Proactivity with Data

With emerging Internet of Things (IoT) technology, vehicles are becoming vast data aggregators, collecting information about various vehicle components in real-time. When combined with blockchain, it forms a powerful alliance.

In an IoT-blockchain system, the data collected by a vehicle is stored on a blockchain, safe from tampering or fraud. Now imagine, specific component data is linked to the manufacturer and the supplier. This insight enables predictive maintenance. When a component appears likely to fail from the data trends, the information is shared across the blockchain to manufacturers and suppliers. The result? Proactive response to potential failures, saving the automotive sector significant costs and the customer unnecessary hassle.

Further down the line, this predictive information could potentially be shared with secondary suppliers, acting as a trigger to manufacture and distribute the necessary components in anticipation of an increase in demand.

6.5. Clearing the Windshield: Transparency and Trust

Transparency is an intrinsic quality of blockchain technology. By making each link in the supply chain visible and traceable, it engenders trust among all parties. Consumers, for one, can authenticate the entire lifecycle of their vehicle, from sourcing of raw materials to final assembly.

The aftermath of the 2015 Volkswagen emissions scandal is palpable evidence. In a post-scandal survey, 59% of respondents reported decreasing trust in all car manufacturers. Implementing a blockchain solution in such scenario would have upheld transparency, possibly restoring, if not enhancing consumers' trust.

Blockchain can also assist in compliance, reducing the friction and costs associated with the verification of environmental standards, safety regulations, and other requirements.

In sum, the application of blockchain technology in the evolution of supply chains is about to shift automotive industry into high gear. It is not just changing the game; it is about creating a whole new game. Buckle up, and let's brave this thrilling new world together!

Chapter 7. Driving Authenticity: Blockchain in Vehicle History Verification

Blockchain technology, originally designed to support the digital currency Bitcoin, is booming in a variety of industries, including the automotive sector. Its role in securing and verifying vehicle history records cannot be overemphasized. A vehicle's history plays a crucial part in its overall value, influencing decisions about buying, selling, and insuring vehicles. However, existing systems often lend themselves to falsifications, manipulations, and omissions, with consumers bearing the brunt of these transgressions. Blockchain technology can prevent these issues, effectively bringing authenticity to vehicle history verification.

7.1. The Significance of Vehicle History in the Automotive Industry

The primary function of a vehicle history report is to give a detailed account of a vehicle's past. This includes information about previous owners, service history, accident history, and odometer readings. It also details whether the vehicle has been subjected to any significant damage or not, such as natural disasters or total loss events. While this information is crucial in most vehicle transactions, its authenticity can sometimes be compromised, thereby impacting stakeholders negatively.

A decidedly transformative solution can be found in blockchain technology. The comprehensive distributed ledger system holds potential to curb manipulation, safeguard information, and provide a single indisputable source of vehicle data.

7.2. Blockchain: An Overview

Blockchain is a decentralized, resilient, and transparent database maintained by various participants known as nodes. Each transaction recorded on a blockchain creates a block. Every block is linked to previous and future blocks, forming a chronological chain. The distributed and immutable nature of this technology ensures every participant has the same version of the truth.

The architecture of blockchain technology makes every node a custodian of the entire ledger, promoting transparency and reducing instances of fraud and misrepresentation. With its core properties of immutability and transparency, blockchain is aptly poised to drive authenticity in vehicle history verification.

7.3. Application of Blockchain in Vehicle History Verification

By moving vehicle history records onto a blockchain-based platform, every service event, ownership change, accident, or repair could be recorded in a transparent, tamper-proof manner.

The implementation would begin with the creation of a unique digital identity for each vehicle. This identity would be preserved and unchanged, with each subsequent event or action taken on the vehicle appended to the record.

A typical blockchain-based vehicle history report might include:

- Original manufacturing information: This offers data about equipment packages, color, and manufacturing date.

- Ownership changes: Documenting the timeline of ownership, including each owner's identity and the purchase or sale price.

- Service history: This enlists all scheduled maintenance and ad-hoc repairs, along with detailed information about the service provider.

- Accidental history: Information concerning any accidents the vehicle has been in is recorded.

- Recall information: Recall status and outstanding recall details are tracked in real-time.

7.4. Real-World Examples of Blockchain in Action

Several progressive companies have already begun implementing blockchain technology for vehicle history verification.

For instance, the start-up company, CarVertical, has set itself the task of creating a global and tamper-proof vehicle history registry. The public blockchain incorporates various sources of vehicle information, ensuring the buyers of automobiles have access to reliable and secure data.

Furthermore, MOBI, the Mobility Open Blockchain Initiative, is developing standards to digitize vehicle identities. Their 'VID II Working Group' is pioneering standards for trustworthy digital identities for vehicles.

7.5. The Way Forward

As we delve further into the realm of blockchain-integrated automotive industry, external IoT devices, sensors, and DSRC (Dedicated Short Range Communications) can be further integrated with the blockchain to provide real-time data gathering and updates for each vehicle. This would lead to a truly comprehensive vehicle lifecycle record, paving the way forward for an even more trustworthy and transparent industry.

As blockchain technology continues to improve, the potential applications for vehicle history verification will only expand. Every player in the value chain, from manufacturers to dealers to consumers, stands to gain from a blockchain-enabled, authentic vehicle history verification process.

While challenges like data privacy, cyber-attacks, and scalability still need to be addressed, the potential benefits of blockchain technology greatly outweigh the risks. Indeed, the road to automotive authenticity, though intricate, is clearly charted. As pioneers adopt and adapt, blockchain can revolutionize the vehicle history verification process, driving trust, transparency, and authenticity in the automotive industry.

Our journey into the blockchain-enabled automotive future, though slightly formidable, would corroborate how technology can play an integral part in creating a highway of authentic and verifiable vehicle history, steering us securely into the future of the industry. The exhaust roar of a conventional vehicle data verification is morphing into the quiet hum of a blockchain-powered revolution, and it's time we paid heed. Brace yourself as we embark on this exciting trajectory; the ride promises to be nothing short of revolutionary.

Chapter 8. Fueling Transparency: Payment Systems & Leasing

In an industry grounded in tangible products and physical assets, the concept of revamping the financial pipelines seems a daunting task. Yet, there lies vast potential for the integration of blockchain technology in the areas of payment systems and leasing procedures. The capabilities of this novel technology whisper the promise of a new era of transparency, efficiency, and security.

8.1. Blockchain and Payment Systems in Automotive

At its core, blockchain is a distributed ledger system that allows a wide range of financial transactions to take place securely and transparently. The simplest and most common example of its application in the financial realm is cryptocurrency, specifically Bitcoin. Imagine if such a system is adopted for payment transactions in the automotive industry; the possibilities are immense and transformative.

Traditional payment mechanisms in the automotive transactions are often marred by slow, cumbersome procedures, and susceptibility to fraud. With blockchain, payments can take place instantly, providing real-time clearance and substantially reducing default risk. Thanks to the cryptographic nature of blockchain, transactions are tamper-proof and fraud-resistant.

Moreover, blockchain brings a sense of clarity to the dealership financing and buyback processes. Today, these operations involve several parties including the dealership, OEM, any intermediaries, and the final customer. Each party maintains separate records and reconciliation becomes a complex process leading to delays and disputes. A single, authoritative blockchain ledger reduces these inefficiencies, enabling all parties to view, update and approve transactions in real-time.

8.2. Blockchain Revolutionizing Leasing

Leasing is another complicated and often friction-filled process in the automotive world. It generally includes multiple parties – lease giver, lease receiver, financial institution, and service providers – each with conflicting goals and incentives. Coordinating between all these components and establishing trust is a challenging endeavor.

Blockchain has the potential to transform this space by bringing unparalleled transparency and trust. Lease contracts can be digitized and stored on blockchain, making them tamper-proof and easily understandable. Revolutionary concept of 'Smart Contracts', scripts that automate performance of contractual obligations, can automate the execution of leasing terms such as penalty for excessive wear and tear or automatically executing payment on due dates. The terms of agreement are visible and consistent for all parties.

Moreover, blockchains can reduce the document-related complexities associated with the end-of-lease process. Key details such as mileage tracking, servicing records, and any damage reporting can be securely stored on the blockchain, eliminating the chance of discrepancies and disagreements.

8.3. Encouraging Transparency and Efficiency

Blockchain promises a level of transparency rarely seen in automotive transactions. Each transaction is securely recorded and verified on the blockchain, visible for all permitted stakeholders to see. This not only discourages deceitful practices but also encourages clear communication and easier dispute resolution.

In addition to transparency, blockchain also promotes increased efficiency. Many paperwork-related processes, known for being time and resource-intensive, can be digitized and automated through blockchain technology. This enhances overall productivity and provides significant cost-saving opportunities for companies.

8.4. Final Thoughts

Indeed, the advent of blockchain has brought forth a whirlwind of changes and possibilities for the automotive industry. As the industry rides the wave of digital transformation, blockchain serves as the beacon, directing the sector towards a more transparent and efficient future. The combined adoption of blockchain in payment systems and leasing procedures has the potential to revolutionize automotive financial transactions, reducing inefficiencies, fraud, and disputes along the way.

The road ahead is lengthy and filled with challenges. The exploration and implementation of blockchain technology will not be a straightforward path. Yet, the benefits and potential impact cannot be overlooked or understated. The journey may be long, but the destination holds the promise of an automotive world streamlined and enhanced by the power of blockchain. Buckle up – it's time to push boundaries and propel automotives into the digital age.

Chapter 9. Road of the Future: Blockchain in Traffic Management and Autonomous Vehicles

As we turn an eye to the future, it becomes clear that the highways we've navigated for so long are no longer mere ribbons of concrete and asphalt. Instead, they have been reimagined into dynamic ecosystems that weave together the inherent elegance of blockchain with the power of autonomous vehicles.

9.1. Integrated Traffic Management Systems

First, let's embark on an explorative trip into the landscape of integrated traffic management systems graced by the presence of blockchain. Blockchain, seen from the perspective of dexterity in ledger management and security quotient, has the potential to redefine traffic management. In a blockchain-based traffic system, cars would be able to communicate with one another directly and securely, using a distributive ledger technology to share information in a tamper-proof way.

Smart contracts, the self-executing contracts with the terms of the agreement being written into code, would be instrumental in these systems. Cars could activate smart contracts automatically when certain conditions are met, such as a vehicle wanting to merge into a lane. The cars could communicate, verify the conditions, and execute the merging contract in real time without the need for human interference.

9.2. The Role of Decentralized Infrastructure

Before delving deeper into blockchain's use cases in traffic management and autonomous vehicles, it is crucial to understand the role of decentralized infrastructure in these areas. The main charm of blockchain is that it is decentralized, meaning the control of the network is not in the hands of a single authority.

This is immensely beneficial in traffic management as it fosters faster, more secure communication between vehicles. Another beneficial offshoot of decentralization is the natural resilience to tampering and fraud. The integrity of the data network could thus be ensured. Odometer fraud, a significant issue that results in massive annual losses, could be mitigated with the tamper-proof nature of blockchain ledgers.

Decentralization could also be a driving force in garnering public participation on a huge scale. People could be incentivized for traffic rule compliance by rewarding them with tokens, which could further lead to less traffic violations.

9.3. Exploring Blockchain's Potential

Now that we have an understanding of the role of smart contracts and decentralization, we can start to look at the specific use cases of blockchain in the realm of traffic management and autonomous vehicles.

Blockchain can be an excellent facilitator in handling parking challenges in congested areas. With a decentralized parking management system in place, cars could communicate with each other about available parking spaces. The parking slot could then be reserved with the help of smart contracts. Once the vehicle leaves the parking spot, the smart contract terminates, making the space available again for other cars.

In the case of autonomous vehicles, blockchain could be used in decentralizing data storage. Large car manufacturers currently maintain data centers for their self-driving vehicles. Every car collects and processes enormous data during its operation and sends it back to the data center. This centralized storage, however, raises concerns about a single point of failure, data manipulation, and cybersecurity threats.

Blockchain could distribute the data and thus mitigate these risks, enhancing the safety of the vehicle and its passengers. Moreover, every vehicle could have a digital identity on the blockchain, including its maintenance history, previous owners, and accident reports. Buyers could verify the car's history before purchase, ensuring full transparency and trust.

9.4. Roadmap to Autonomous Vehicles

Now let's delve into the future with autonomous vehicles on blockchain-enabled roads. Self-driving car technology is an increasing reality and experts believe blockchain could facilitate this transition.

On these blockchain enabled roads, vehicles will have their own digital identity. All their actions and transactions will be recorded on the blockchain, ensuring traceability and accountability.

The journey of an autonomous car would start by selecting the fastest route based on real-time traffic data from other vehicles and infrastructure. This information sharing will be facilitated by blockchain, ensuring secure and tamper-proof data exchange.

Next, when the car reaches a toll booth, it won't need to slow down, but rather, a smart contract will be activated, making the transaction smooth and efficient. The car could also pay and reserve a parking spot for itself with the help of a smart contract.

9.5. In conclusion

Blockchain stands as a stalwart of robust capabilities and utmost security, waiting to be adopted by the transport sector. With blockchain, trips will be faster, more efficient, secure, and environmentally friendly. Congestion and pollution will drastically reduce due to intelligent traffic management, facilitated by decentralization and automation.

As we look towards this promising horizon, it remains essential to understand that the adoption of these technologies will not be without challenges, including the difficulty of completely altering the conventional systems, addressing the concerns of data privacy, and the legal complications tied with autonomous vehicles.

However, as the shift in the automotive industry accelerates towards smarter and more efficient systems, the blockchain technology encapsulates an extraordinary number of benefits that are enticing for manufacturers, drivers, and the overall infrastructure. The road of the future, it seems, is paved with blocks of code.

Chapter 10. Bumps Ahead: Challenges in Implementing Blockchain

Before diving headlong into the potential benefits of blockchain in the automotive sector, it is crucial to address the imposing challenges that lay ahead. Implementing a powerful, complex tool like blockchain in an industry as multifaceted as automotive is no leisurely drive in the park. It's more akin to maneuvering through a winding mountain road in the middle of a thunderstorm. Suffice it to say, there are bumps on this road.

10.1. Clearing the Fog: Understanding Blockchain

Blockchain technology, an ingenious solution that is credited to a person (or group of people) known as Satoshi Nakamoto, was originally conceptualized to establish the digital currency, Bitcoin, to combat double-spending without the need of a central server. At its core, blockchain can be visualized as a transparent, verifiable, decentralized ledger of all transactions across a peer-to-peer network. While simplicity is the primary essence of its appeal, the reality of understanding the technology is far from straightforward.

Blockchain is a subtlety nested paradox: simple in its concept, yet sophisticated in its operation. Gaining enough understanding to not only integrate it into operations, but also to innovate, necessitates extensive investment in time and resources, which is a crucial hurdle. Most decision-makers still struggle to understand its core principles. Add to that the fact that the technology is continually evolving and thus, needs recurring educational investment makes mastering blockchain quite challenging.

10.2. Trust: The Invisible Battleground

While blockchain was created to circumnavigate the need for trust in financial transactions, ironically it now faces the challenge of gaining that trust from industries it wishes to revolutionize. Acceptance of any new technology is deeply intertwined with trust, and blockchain is not an exception to this rule. Despite its potential transformative impact on the automotive sector, the fact that blockchain is tethered to Bitcoin, which is often associated with volatility, speculation, and illicit activities, adds an unnecessary layer of mistrust.

Overcoming this trust barrier isn't just a matter about dispelling myths around the technology or decoupling its perception from cryptocurrencies, but also about convincing stakeholders of its security. The consensus mechanisms and encryption techniques ensure security, but the looming threat of quantum computing and hackers finding weak spots can undermine the integrity of the system at any given point.

10.3. Regulatory Roadblocks and Standardization Jams

Establishing and enforcing globally uniform standards and norms are essential steps to facilitate the implementation of blockchain technology in the automotive industry. Without standardization, the promise of interoperability, seamless data transfer, and robust security that blockchain offers is like having a vehicle without wheels.

Moreover, legal and regulatory hurdles prove to be another vexatious obstacle. Every country has its own data privacy regulations and laws, which can hinder the smooth implementation of blockchain. Regulations around data management, like GDPR in

Europe, may muddle the implementation due to compliance requirements. The lack of clear legal treatment of blockchain technology also contributes to the hesitance in adopting it.

10.4. Crumbling Infrastructure Worries

Leaping ahead to a blockchain-driven system requires a robust infrastructure that can handle high volumes of data with lightning-speed. Yet, many countries, particularly developing ones, suffer from limited or unreliable network connectivity. Furthermore, the IT infrastructure in many corporations might not be equipped to accommodate the strain blockchain systems might place on them.

Another infrastructure setback is the massive energy consumption. Running Blockchains involves complex computations which require a lot of computing power, resulting in high energy consumption. This additional demand could put stress on power grids, making it a potential environmental concern, especially amidst global efforts to reduce carbon footprints.

10.5. Skills Shortage: A Speed Limit?

Blockchain, being a relatively new technology, faces a deficit of skilled professionals able to integrate it into existing systems and develop innovative applications for automotive. A limited talent pool can severely affect the speed of migration, increase costs, and create a gap of knowledge. Furthermore, the competition for these limited experts is fierce, often inflating their worth and thereby, potentially making blockchain implementation infeasible for smaller companies or those with limited budgets.

In conclusion, the path to blockchain's successful implementation in the automotive industry isn't challenge-free. From clearing the fog surrounding it, establishing trust, navigating regulatory roadblocks and standardization jams, to tackling crumbling infrastructure worries and countering the skills shortage make up a formidable array of bumps. However, history has provided ample evidence that technological evolution is an unstoppable force. Therefore, it is safe to presume that the automotive sector will eventually find its way to smoothly implement blockchain, optimizing the industry in phenomenal ways never seen before. Like any journey, it would require a firm stepping foot, patience, and unwavering commitment.

Chapter 11. Looking Through the Windshield: The Future of Blockchain in the Automotive Industry

Blockchain technology, an offspring of the digital age, is poised to inject unprecedented levels of transparency, efficiency, and security into the automotive industry. While traditionally associated with financial transactions, the potential applications of this technology go far beyond cryptocurrency, particularly in shaping the future of the automotive and transportation sectors.

11.1. The Underlying Technology: A Simplified Overview

Before delving into applications, it's crucial to understand the mechanics of blockchain technology. At its heart, a blockchain is just a public, decentralized, and digital ledger of any transactions in any type of record. Information, once added to the chain, cannot be altered or tampered.

Each 'block' contains a list of transactions. These blocks are linked together using cryptographic principles, forming a 'chain'. This technology's most potent trait is its transparency: everyone on the network can view transaction history. Contrary to popular belief, this transparency doesn't sacrifice privacy, as real identities of individuals aren't disclosed.

11.2. Blockchain in Automotive: Current Applications

So, how does a technology primarily known for its use in the financial sector translate to the automotive industry? Let's examine the current applications:

1. The Supply Chain: Blockchain's decentralization and traceability capabilities make it ideal for managing supply chain complexities. It enables real-time tracking of parts' veracity, which could drastically reduce instances of counterfeiting and ensure only quality parts are adhered to.

2. Car Leasing and Sales: Vehicle leasing and sales involve a consortium of participants - finance providers, dealers, insurers, etc. Here, blockchain simplifies the process by reducing paper-based trails, therefore lessening chances of fraud while boosting efficiency.

3. Automotive Financing: Blockchain-based smart contracts can automate many aspects of automotive financing, from loan origination to payments, which could make the entire process more transparent and efficient.

11.3. The Road Ahead: Future Applications

While current uses are impressive, the potential future uses are where the excitement truly lies.

1. Autonomous Vehicles: As self-driving cars become more prevalent, blockchain could play a critical role in data management, security, ride-sharing platforms, and even in creating a universal transport protocol. It can ensure that data from vehicles, such as real-time traffic updates or meter readings,

is reliable and secure.

2. Vehicle Identity Solutions: With blockchain, a 'digital passport' for each vehicle can be created, documenting every detail from manufacture to end-of-life.

3. Traffic Management and Smart Cities: Traffic congestion is an all-too-familiar problem for metropolitan areas. Blockchain can enable a decentralized, peer-to-peer system of communication between vehicles, traffic infrastructure, and pedestrians, thereby optimizing traffic flow in real time.

4. Intelligent Insurance: Blockchain can also transform the insurance industry through usage-based insurance schemes. Here, individual driving behavior data can be stored securely and used to determine premiums, thereby making the system more personalized and fair.

11.4. Challenges and Framework for Implementation

While blockchain presents immense potential, implementing it isn't without its challenges. Some of these include scalability, energy consumption, data privacy, and regulatory uncertainties.

Regardless of these hurdles, it is clear that blockchain is key to unlocking a more efficient, transparent, and secure future for the automotive industry. The way forward, therefore, is not to resist but to prepare for inevitable change. First, understanding the technology itself and its potential applications is crucial. Next, capabilities need to be developed through experimentation and collaboration, fostering a culture of innovation. Finally, setting up governing standards that balance the interests of all participants in the ecosystem will be paramount to ensuring sustained and mutually beneficial growth.

In conclusion, the fusion of blockchain technology with the

automotive industry is a potent one that promises to revolutionize various aspects of the industry. It's a thrilling journey that's just starting, and it's one that we should all join, not just for the promise of astounding transformations but also for its undeniable potential to redefine our future. The course ahead, indeed, looks promising and full of potential new avenues for exploration. So buckle up as the automotive industry shifts gears on this exciting technological road trip.

www.ingramcontent.com/pod-product-compliance
Lightning Source LLC
Chambersburg PA
CBHW072219290526
45794CB00007B/2815